WORLD WITHIN A WORLD

Pribilofs

Books by Ted Lewin

WORLD WITHIN A WORLD

    EVERGLADES

    BAJA

    PRIBILOFS

WORLD WITHIN A WORLD

Pribilofs

Written and illustrated by TED LEWIN

DODD, MEAD & COMPANY • NEW YORK

1 2 3 4 5 6 7 8 9 10

Library of Congress Cataloging in Publication Data

Lewin, Ted.
    World within a world — Pribilofs.

    Includes index.
    SUMMARY: Describes some of the animals to be
seen in the Pribilof Islands, with particular
emphasis on the seals.
    1. Zoology — Alaska — Pribilof Islands.
2. Northern fur seal. [1. Zoology — Alaska —
Pribilof Islands. 2. Northern fur seal.
3. Seals (Animals) 4. Pribilof Islands]
I. Title.
QL161.L48   591.9798'4   80-1009
ISBN 0-396-07855-9

For Betsy

Contents

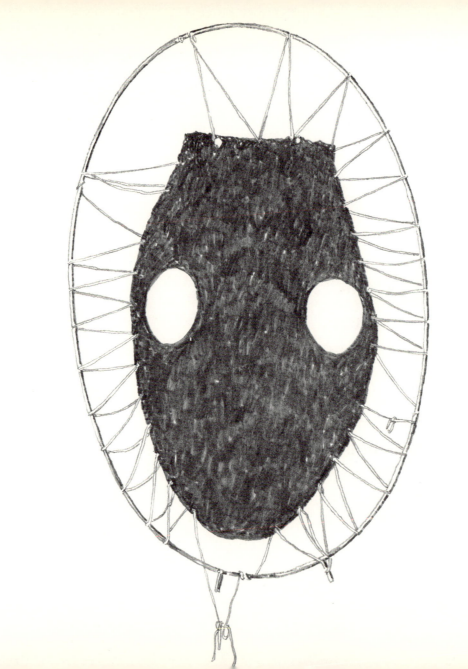

The Islands

The Pribilof Islands are isolated bits of volcanic rock and tundra located off the coast of Alaska in the midst of the vast Bering Sea. Near the top of the world, they are inhabited seasonally by over a million fur seals, countless seabirds, and — only recently — by man.

Man's short history on the islands is one of plunder that saw the seal herd diminish to the point of extinction. The seals were killed for their valuable skins by the Aleut Indians who were brought there as slaves by the Russians nearly two centuries ago. The Aleuts fared badly, as did their seal victims.

The islands now belong to the United States, and the herd has been "managed" back to 1.3 million animals. The Aleuts still live there and continue to harvest the seals.

The herd is more or less stable at its present size; however, no one knows how much longer it can continue to bear the pressure of the killing of 25,000 of its number each year.

I arrived in the Pribilofs late one August. The killing was over, and the seal rookery was going about its business.

Thousands of seabirds soared and hovered around their cliff homes. I wandered over Alpine tundra in search of any tiny wildflowers that might remain from summer profusion, and I felt the cold emptiness of the Bering Sea crashing against the fog-shrouded islands as if to rid itself of such tiny intrusions on its surface.

The Rookeries

Each spring the tundra is not yet free of snow when the first big, breeding males — the beachmasters, heavy with blubber — pull themselves from the sea onto the mist-shrouded shores. Each establishes his own territory, and any other adult males enter at considerable risk.

In mid-June the first females arrive. They are swollen with full-term pups. As they head for a place on the beach, they are intercepted by the nearest males. Try as he may, a beachmaster cannot hold a female if she has another site selected. The choice is ultimately hers.

Each bull finally will collect between one and one hundred females.

The females give birth within forty-eight hours of coming ashore, and will mate again within a week. Other than nursing, very little maternal care is given the pups.

Beachmasters never rest from their endless fighting and herding. The air is filled with their rasping, clicking sounds, and roaring and bleating. Savage fights break out between rival males. They bite each other, sometimes holding on for hours, letting go only for a better purchase. If a female should try to slip out of his territory, the beachmaster will seize her, often throwing her high into the air and back into his harem.

Young sub-adult males not yet able to compete with the beachmasters haul up on nearby beaches. They wait their chance and, in late August, will try to work a spot on one of the rookeries.

Pups are born with a thin coat of black hair. They are at the mercy of the elements until their first molt in October, when they acquire a coat of thick silver fur. In the short breeding season, over 360,000 pups are born, or one every ten seconds. It's fortunate that Nature is so bountiful, as great numbers die in their first year.

The frenetic activity of the rookeries continues unabated all during June and July.

16

Now, in August, standing on the edge of the tundra, I look down into the black amphitheatre of the beach. The scene is chaotic. Thousands of seals are swimming offshore, boiling the surface like schools of bait fish. Thousands more jam the beach. Bleats, groans, roars, and throaty gargles are borne on the cool mist. The beachmasters are spent and lethargic. They have not eaten for over two months.

18

The pups are growing stronger, and they wander around the rookery in pods, bleating like lambs.

One idle bull works his way up the gradual slope of the beach and asserts himself with two young females. The nearby beachmaster rouses himself from his torpor. He is gaunt, bone weary, and used up. His deep-reddish skin hangs in great loose folds. Slavering, whiskers erect, he looks at the intruder. The young bull seems to want a test of his maleness. The beachmaster, challenged, makes a lightning charge. A pod of terrified pups scatters from the onslaught. The old bull is awesome even in his weakened condition. The young bull is no match and quickly takes refuge behind a large rock. Peace is momentarily restored to this tiny portion of the rookery.

A soaking wet female appears high up on the beach, looking for her pup. She may have been two hundred miles at sea for a week or more, feeding, but will find her own pup among all of these thousands. She climbs on a rock and cries out for him. . . .

24

She touches noses and sniffs a pup . . . no, not hers.
Her search continues. None in this pod of wet little pups
is hers. . . .

She scratches prodigiously,...

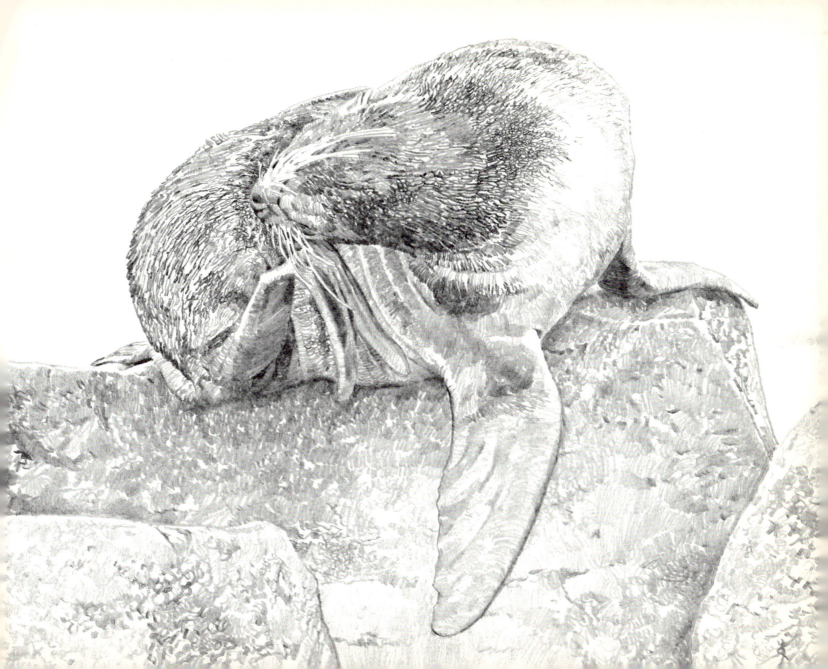

and naps for a while. It has been a long swim.
When she awakes, she resumes her
search. The touched noses finally match, . . .

and she rolls over on her side, lifting a hind flipper so her hungry pup can find the source of her rich milk.

By the end of August the beachmasters will have left the Islands to begin their hard winter at sea — no one knows where. Soon after the bulls leave, the pups will be fat and ready for their long swim. One day in the coming October or November, their mothers will not return. The pups will line the beach, looking anxiously out to sea. Finally they will plunge into the cold water for a swim that, for some, may last as long as two years before they return to these shores. Then, once again, the beach, scene of so much violence and the miracle of so much new life, will be empty and silent.

The Bird Cliffs

August. Dense, wet carpets of cow parsnip and beach rye cover the tops of the jagged cliffs on the South Cape near the Reef Rookery. Below, murres cling to the rocks like black-and-white lichens. Newcomers hover and veer off, finding no more room on the precarious perches. Tufted puffins flutter by along the cliff's edge. Wheeling out over the crashing surf far below, the puffins let the wind carry them back around the point.

40

In the opaque blue water, seals feed alone or in small groups. Some move very close to the cliffs, seemingly unaware of the danger from crashing waves. Some ride the swells, and the dark shapes of their bodies can be seen below the surface of the water.

The wind carries the acrid odor of the thousands of birds out of sight on the sheer cliffs below me. Rain falls heavily now. More and more puffins pass and wheel off. Across the windswept tundra behind me, on the opposite shore of the island, are the masses of seals on their dark beach. Faintly, I can hear their grunts and bleats.

The 400-foot cliffs of Southwest Point are softened by green ledges of faery flowers. Murres in crowds stand with their backs to the sea. . . .

Holding on for dear life, I peer over the edge for a better look at the murres. They are safe and secure in their rocky homes.

In and out from their cliff dwellings, horned puffins with clean, clown faces hover and land. Their colorful, parrot-like bills make yellow and red splashes against the soft green vegetation and dark gray rock. One carefully and thoughtfully preens himself. He wags his head comically from side to side and settles down to rest.

50

Fulmars squawk at each other.

Kittiwakes, horned puffins, and murres dive and wheel, the kittiwakes with bills full of fresh nesting material.

I see a kittiwake chick. It stretches its scrawny neck and pecks at its mother's bill. Below, the sea crashes onto the black, bouldered shore, setting off the immaculate white bellies of the glaucous-winged gulls as they soar and lift with the thermals.

Rafts of murres ride the swells atop the kelp. One chick, a tiny replica of its mother, is there with them. Too young to be in the water, it must have fallen from the nest. It probably won't survive to fly with the rest for its winter at sea.

56

The Tundra

A raven appears out of the mist and lands atop the lichen- and moss-covered chunks of lava that, like the crumbling monoliths of a people long gone, sit heavily on the fog-shrouded tundra. Slowly disintegrating, the lava chunks too will become tundra with time.

The tripods and catwalks used by the scientists to count the seals for harvest take on the delicate feel of etched lines on a zinc plate. They are strangely right with the natural lines of sea, rocky shore, and dunes; their harmonious forms belie their ominous function....

58

Wet winds bearing the strong musk of a million fur seals...

Tiny Martian gardens in teacup-sized depressions in the lava...

An arctic fox comes down to Big Lake for a drink. He stands shoulder deep in the cow parsnip and brilliant yellow beach fleabane. Suddenly frightened, he is a gray shadow running lightly through the wet beach rye. He disappears over the dune. A gray mist returns...

62

A horizon of beach rye with a head and two gray ears.
The fox barks...and melts into the tundra. Above its
burrow, a patch of gray sand...on the sand, the remains of
a seal pup. I recall a stretch of beach littered with dead
seal pups, glaucous-winged gulls, dressed disrespectfully
in white, picking them over.

A salt pond is surrounded by the golden copper of
marsh grass. Two female pintail ducks sit on the water,
their blue bills the palest accent of color.

66

The groans and bleats of thousands of fur seals shift
and fade into the thick mist...and disappear in the
vacuum of the cold Bering Sea, leaving a primal feeling of
apprehension.

One lone, dark female seal on a field of tundra
flowers, softened by the gray sea mist....

68

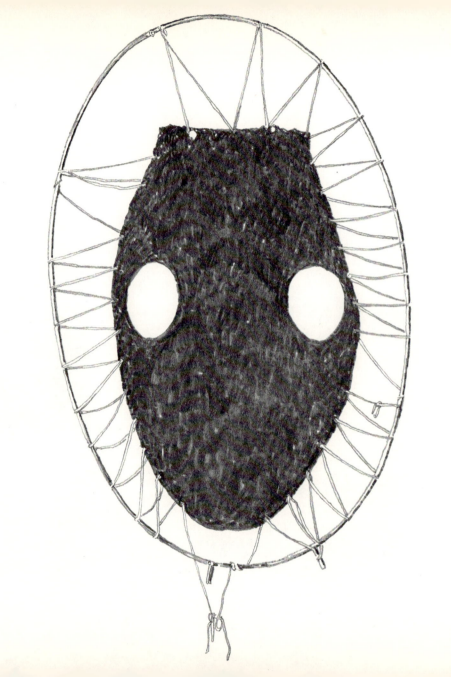

The Future

Until 1911, when a treaty was signed by Russia, Canada, Japan, and the United States, the fur seals were being hunted to near extinction in the open sea. Out of this treaty grew the annual "harvest" on the islands.

As the end of the four-nation treaty approached, there was mounting pressure from some to let the treaty lapse with a return to hunting on the open sea. Others felt that there was no justification for killing the seals, economic or otherwise, and wanted them protected, and the Pribilofs to become a sanctuary, with no killing on land or sea.

The fate of these fascinating creatures is yet to be determined.

LIST OF ILLUSTRATIONS

76

INDEX

ABOUT THE AUTHOR-ARTIST

TED LEWIN was born and brought up in Buffalo, New York. He attended Pratt Institute of Art in Brooklyn, New York, where he and his wife now live in a hundred-year-old brownstone.

While his art shows versatility in both style and subject matter, his main interest has always been the natural world and its inhabitants.

Ted Lewin is the author of two other titles in the World Within a World series, *World Within a World — Everglades*, and *World Within a World — Baja*.